Nazim Baluch

Innovation Imperative

AF138520

Nazim Baluch

Innovation Imperative

Actualizing Ingenious Ideas at a Canadian Auto Metal Stamping Plant

LAP LAMBERT Academic Publishing

Impressum / Imprint

Bibliografische Information der Deutschen Nationalbibliothek: Die Deutsche Nationalbibliothek verzeichnet diese Publikation in der Deutschen Nationalbibliografie; detaillierte bibliografische Daten sind im Internet über http://dnb.d-nb.de abrufbar.
Alle in diesem Buch genannten Marken und Produktnamen unterliegen warenzeichen-, marken- oder patentrechtlichem Schutz bzw. sind Warenzeichen oder eingetragene Warenzeichen der jeweiligen Inhaber. Die Wiedergabe von Marken, Produktnamen, Gebrauchsnamen, Handelsnamen, Warenbezeichnungen u.s.w. in diesem Werk berechtigt auch ohne besondere Kennzeichnung nicht zu der Annahme, dass solche Namen im Sinne der Warenzeichen- und Markenschutzgesetzgebung als frei zu betrachten wären und daher von jedermann benutzt werden dürften.

Bibliographic information published by the Deutsche Nationalbibliothek: The Deutsche Nationalbibliothek lists this publication in the Deutsche Nationalbibliografie; detailed bibliographic data are available in the Internet at http://dnb.d-nb.de.
Any brand names and product names mentioned in this book are subject to trademark, brand or patent protection and are trademarks or registered trademarks of their respective holders. The use of brand names, product names, common names, trade names, product descriptions etc. even without a particular marking in this works is in no way to be construed to mean that such names may be regarded as unrestricted in respect of trademark and brand protection legislation and could thus be used by anyone.

Coverbild / Cover image: www.ingimage.com

Verlag / Publisher:
LAP LAMBERT Academic Publishing
ist ein Imprint der / is a trademark of
OmniScriptum GmbH & Co. KG
Heinrich-Böcking-Str. 6-8, 66121 Saarbrücken, Deutschland / Germany
Email: info@lap-publishing.com

Herstellung: siehe letzte Seite /
Printed at: see last page
ISBN: 978-3-659-57853-3

DEDICATION

This book and my love are dedicated to Fadia, my wife and best friend, who

lives in my heart, forever.

CONTENTS

2

SUMMARY

In industrialized countries manufacturing firms are facing significant change resulting from mass customization, shortening product life cycles, increasing technological change, and the entry of international competitors into their markets as witnessed by the automotive and electronics industries. Innovation is the creation of better or more effective products, processes, services, technologies, or ideas that are accepted by markets, governments, and society. It involves deliberate application of information, imagination, and initiative in deriving greater or different value from resources, and encompasses all processes by which new ideas are generated and converted into useful products. 'Process Innovation' is one of the key areas where innovation is exigent. It pertains to finding better or more efficient ways of producing existing products, or delivering existing services. Making innovation a ubiquitous capability in manufacturing is fundamentally a leadership challenge. It needs a tangible organizational infrastructure that makes managers accountable at all levels for driving, facilitating, and embedding the innovation process into every part of the culture. Creativity is a basic human capability. Employees have ideas regardless of whether or not the environment is conducive but the employee will not submit them if the environment is not seen as supportive. Creativity can be encouraged, enhanced and enabled by managers. Although so much innovation today emerges through 'group' processes, most literature on creativity focusses on creative individuals. Any group can be more creative, even its members individually wouldn't score highly on tests for creativity. This book construes the process of 'impelling and managing innovation' at an automotive metal stamping plant in Canada, by Beta plant's dedicated team, during the 2008-2009 financial crises in North America. The book discusses the processes, psychological and physical environment, organisational culture, economic climate, and program content; rationale, significance and denouement. The book concludes that idea management systems don't replace traditional departments and processes involved in new services, products, or strategies. They serve as an adjunct to them and provide a framework that can help organizations turn innovation into an enterprise-wide discipline-and a sustainable process that drives growth in good times and bad. Auto industry is going to remain important, for local, regional and national economies, as well as for the future of the planet if ecologically sustainable transport systems are to be developed; it will, no doubt, remain an important topic in academia.

4

1. BACKDROP

Looking back over the 20th century, one may regard the auto industry as a metaphor for capitalist development. The 20th century was dominated by the development and roll-out, globally, of mass production and consumption, described as the 'first revolution' in auto production; with large factories and Taylorist regulation of assembly-line speeds and techniques. That whole era, covering the first 80 years of the century, was often referred to as the era of 'Fordism', after Henry Ford's first production-line factory (Womack *et al.,* 1990). Much of this 'post-Fordist' literature was rather superficial, not least in its parochial view that developments in the industrialized economies typified a global development, whereas if anything the opposite was the case. To the extent that the large factories in Western Europe and North America were giving way to smaller scale facilities and service provision, this was mirrored by the rise of mass production in the less developed countries (Costello *et al.*, 1989). In terms of production technologies, post-Fordism refers to the sort of lean production, just in time processes applied in Japanese industry, and in particular, Toyota production systems: again, it was indeed an auto company that appeared to exemplify this socio-historical–geographic shift, or 'second revolution' (Womack *et al.,* 1990). Since the mid-1990s, a so-called 'third revolution' has centred on improvements in flexibility, with implications for product creation, design, and manufacturing and life cycle (David *et al.,* 2010).

Though the Original Equipment Manufacturers (OEMs) currently prefer to locate near the final market, they have shifted assembly operations towards low-cost locations within major trade blocs—towards central and Eastern Europe for example, within the EU. Similarly, in the USA, there has been a shift southwards and to Mexico, with auto production now located primarily in the 'auto-valley' (Klier & Rubenstein, 2010). This was driven by the restructuring of the 'Big

Three' (GM, Ford, and Chrysler) producers and the setting up of transplant plants by foreign firms. The end result has been a division of auto-valley into two subareas, a northern area dominated by the 'Big Three' and a southern area dominated by foreign-owned carmakers (David *et al.,* 2010).

The global financial crisis had dramatic impact on auto manufacturing worldwide. However, these were felt uniquely severely in North America, largely because of its asymmetric position within the geography of automotive globalization. North American automakers were already fragile due to one-way trade and foreign direct investment inflows. This history also shaped the nature of the North American policy response. Unlike other jurisdictions, North American governments needed to save leading regional producers from liquidation. Moreover, this rescue took on a unique anti-union tone, through government-mandated renegotiation of labour contracts. The measures taken in North America, while dramatic, are unlikely to resolve the continental industry's deeper structural weaknesses (Stanford, 2010). Financial crisis can be seen as laying bare a range of pre-existing vulnerabilities in the auto industry. Indeed, even before the global financial crisis unfolded, the 'Big Three' North American OEMs ran up total net losses of over $100 billion between 2005 and 2008, exhausting their equity base and questioning their viability (Stanford, 2010). In addition, the Big Threes' efforts at generating profits before the crisis through 'financialization' strategies—in effect 'auto banks' financing sales, leasing and derivatives operations exposed them to an additional set of risks when faced with a 'double whammy' of shocks on both the financial and industrial side (Freyssenet & Jetin, 2009).

2. MAGNA INTERNATIONAL and BETA AUTO STAMPING PLANT

Magna International is the most diversified automotive supplier in the world; they design, develop and manufacture automotive systems, assemblies, modules and components, and engineer and assemble complete vehicles, primarily for sale to OEMs of cars and light trucks in three geographic segments - North America, Europe, and Rest of World (primarily Asia, South America and Africa). They have 315 manufacturing operations and 89 product development, engineering and sales centers in 29 countries on five continents (Magna, 2013). Though Magna International had a healthy balance sheet and an adequate cash flow, it was still susceptible to financial volatilities that had gripped the automotive sector. Being the preferred 'tier 1' supplier to most auto OEMs, Magna was not immune to afflictions of the prominent OEMs including the 'Big Three'; GM, Ford, and Chrysler.

These were desperate times for the North American automotive sector; with downsizing and plant closures, Magna initiated drastic cuts across the board. Beta plant, though profitable, was also handed over drastic budget cuts and its management was told to do more with less or else face plant closure; some of Beta's equipment was already in transit to the 'auto-valley' in the south. Beta Industries, employing fewer than two hundred, is a small, automotive, metal stamping facility in southern Canada. It is tier-1 supplier of structural metal stampings to auto OEMs. Beta's equipment comprised of eight 400-800 tons capacity mechanical stamping presses and several automated robot assembly units. 1500 ton capacity, mechanical stamping press (referred as 'Press' later in the book) was an exception; this was a newer press added to this 25 year old plant in 2005. In order to accommodate the five stories high Press, plant's roof height had to be raised from 60 to 100 feet and the west side wall had to be knocked down to rig it in. This Press provided sixty percent of Beta plant's revenues.

3. PROGRESSIVE METAL STAMPING

Stamping presses and stamping dies are tools used to produce high volume sheet metal parts. These parts achieve their shape through the effects of the die tooling. Production stamping is generally performed on materials .020 to .080 inch thick, but the process also can be applied to foils as thin as .001 inch or to plate stock with thickness approaching 1.000 inch.

Formability is the primary attribute of sheet metal material and is defined as the materials ability to be; bent; stretched; and drawn. The word 'die' is a generic term used to describe the tooling used to produce stamped parts. A die set assembly consisting of a male and female component is the actual tool that produces the shaped stamping (Figure: 1).

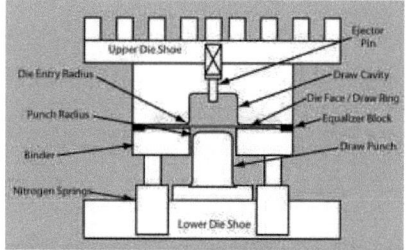

Figure: 1   Simple die assembly

Male and female components work in opposition to both form and punch holes in the stock. The upper half of the die set, which may be either the male or female, is mounted on the press ram and delivers the stroke action. The lower half is attached to an intermediate bolster plate which in turn is secured to the press bed. Guide pins are used to insure alignment between the upper and lower halves of the die set (SME, 2013).

Progressive die drawing, or stamping, is a forming process that utilizes a series of stamping stations to perform simultaneous operations on sheet metal. The final metal work piece is developed as the strip of metal is processed through the stamping die. The progressive die stamping process characteristics include: the utilization of multiple cutting and/or forming operations simultaneously; excellent suitability to produce small work pieces at a rapid rate; the necessity to invest in expensive die sets; the ability to save time and money by combining forming operations; and the capability to maintain close tolerances, depending on the tools. The illustration that follows provides a two-dimensional look at a typical progressive die metal drawing process in two steps – one open die and one closed die. As the metal strip is moved through the drawing process, it is exposed to a series of progressive die stations, each one changing the metal configuration left on the metal by the previous station. Therefore, the metal work piece is created in a series of stamping stages.

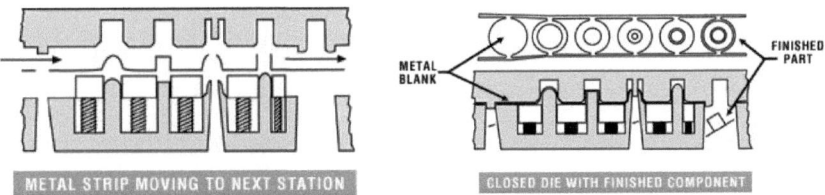

Figure: 2 Open Die (Advantage, 2014)   Figure: 3 Closed Die (Advantage, 2014)

Figure: 2 depicts the metal strip moving to the next workstation before the die closes while Figure: 3 shows a closed die with the finished component work piece falling out of the die on the lower right hand side.

Figure: 4 Progressive Metal Stamping Die (Source: Metal Stamping Services – Chicago USA)

Progressive stamping is a metalworking method that can encompass: Blanking; Punching; Coining; Bending; Perforating; Piercing; Notching; Lancing; Embossing; and several other ways of modifying metal raw material, combined with an automatic feeding system (Figure: 4). The feeding system pushes a strip of metal (as it unrolls from a coil) through all of the stations of a progressive stamping die. Each station performs one or more operations until a finished part is made. The final station is a cut off operation, which separates the finished part from the carrying web. The carrying web, along with metal that is punched away in previous operations, is treated as scrap metal. The progressive stamping die is placed into a reciprocating stamping press. As the press moves up, the top die moves with it, which allows the material to feed. When the press moves down, the die closes and performs the stamping operation. With each stroke of the press, a completed part is removed from the die. Since additional work is done in each "station" of the die, it is important that the strip be advanced very precisely so that it aligns within a few thousandths of an inch as it moves from station to station. Bullet shaped or conical "pilots" enter previously pierced round holes in the strip to assure this alignment since the feeding mechanism usually cannot provide the necessary precision in feed length (Baluch *et al.*, 2012).

4. ADVACED HIGH STRENGTH STEEL and its STAMPING CHALLENGES

The world's most common alloy, steel is the material of choice when it comes to making products as diverse as oil rigs to cars, planes to skyscrapers; simply because of its functionality, adaptability, machine-ability and strength (DOCOL, 2012). To improve crash worthiness and fuel economy, the automotive industry is, increasingly, using Advanced High Strength Steel (AHSS). Today, and in the future, automotive manufacturers must reduce the overall weight of their cars. The most cost-efficient way to do this is with AHSS. However, there are several parameters that decide which of the AHSS types to be used; the most important parameters are derived from the geometrical form of the component and the selection of forming and blanking method (SSAB, 2013). Using AHSS offers many environmental advantages, as well; when weight is reduced in producing a detail, a smaller amount of raw material is used and less energy is consumed. At the same time, less energy is needed to transport the steel (Uddeholm, 2013).

AHSS Benefits: AHSS is most advantageous when used for safety components, structural parts of the car body and the chassis. As a general rule, the weight reduction is about 50 percent, i.e. when compared to mild steel the thickness is halved without sacrificing strength. As a result, AHSS has become the material of choice for structural parts including: sill reinforcements, A-pillars, B-pillars, side impact beams, bumpers, roof bows, seats, and many more (Figure: 5). AHSS steel has also been introduced in all other vital areas of the car and many new cars are already composed of 30 to 40 percent AHSS. In only a few years, AHSS is predicted to make up 40 to 50 percent of the sheet steel used in cars contributing to a 5 percent reduction in total GHG (Green House Gas) emissions.

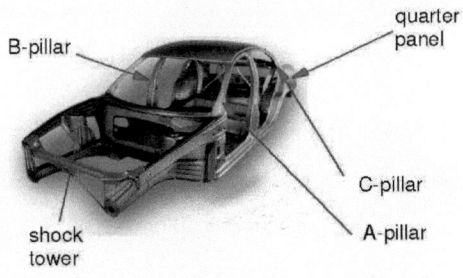

Figure: 5 Car Body Showing Structural Parts

The advantages of using AHSS are evident in all types of cars. Small cars benefit from greatly improved crash safety while heavier cars, like SUVs, can be made much more fuel efficient.

The smooth and cost-efficient transition from mild steels, or conventional high strength steels, to AHSS is one of the factors behind the success of AHSS within the automotive industry. This is because AHSS can be formed and joined in much the same way as milder steel, but with only half the thickness. The cutting and forming of AHSS can be done without increased force and common welding and joining techniques can be used with only minor adjustments.

Even greater benefits can be gained when switching to new forming methods, like roll forming. This process enables the manufacturer to produce complicated profiles in a single run; roll forming can be done with lower tool wear and very tight radii of the profile when using AHSS. The advantages of AHSS over aluminium are easily summarised: lower cost with the same weight and strength. The ability to form complicated profiles is another of the many benefits offered by AHSS compared to those of aluminium; on beams inside the car, aluminium must be three times the thickness of AHSS to provide the same strength. When compared with new materials, such as composites, the cost benefits of using AHSS is even more apparent. The automotive industry of today has decided that AHSS is superior to aluminium for nearly all applications.

Challenges in Forming AHSS: Forming of AHSS is not a radical change from forming conventional High Strength Steel (HSS). The major acquisition of new knowledge and experience needed for forming higher strength steels in general increased gradually over the years as ever-increasing strengths became available in the HSLA (High Strength Low Alloy grades). Now new demands for improved crash performance, while reducing mass and cost, have spawned a new group of steels that improve on the current conventional base of HSS. The improved capabilities the AHSS bring to the automotive industry do not bring new forming problems but certainly accentuate problems already existing with the application of any higher strength steel. These concerns include higher loads on presses and tools, greater energy requirements, and increased need for springback compensation and control. In addition, AHSS have greater tendency to wrinkle due to lack of adequate hold-down and often a reduction in sheet thickness.

Matching exact mechanical properties of the intended steel grade against the critical forming mode in the stamping not only requires an added level of knowledge by steel suppliers and steel users, but also mandates an increased level of communication between them.

New emphasis is being placed on determining specific needs of the stamping, highlighting critical forming modes, and identifying essential mechanical properties. The interaction of all inputs to the forming system means the higher loads and energy needs of AHSS also place new requirements on press capacity, tool construction/protection, lubricant capabilities, process design, and maintenance.

Since many applications of AHSS involve load bearing or crash analyses, computerized forming-process development has special utilization in structural analysis. Previously the part and assembly designs were analysed for static and dynamic capabilities using CAD (Computer-Aided Design) stampings with initial sheet thickness and as-received yield strength. Often the tests results from real parts did not agree with the early analyses because real parts were not analysed. Now virtual parts are generated with point-to-point sheet thickness and strength levels nearly identical to those that will be tested when the physical tooling is constructed. Deficiencies of the virtual parts can be identified and corrected by tool, process, or even part-design before tool construction has even begun.

In addition, the new higher-strength steels require more energy to stamp, which introduces heat into the process. Heat can be controlled by slowing the stamping-line speed, but productivity suffers. Die-cooling methods may have to be introduced to combat the heat. And of course, anything that helps to reduce friction will be a friend to manufacturers who need to stamp parts using these new steels. Drawing, or draw forming, involves forcing a blank deeply into a die cavity and shaping it into the shape and contour of the punch face and sides. Without sufficient formability qualities, drawn blanks are subject to wrinkling, thinning, and fracturing. The resistance of the sheet metal stock to the forces exerted by the moving dies creates friction. For this reason, lubrication is vital for successful sheet metal forming and as the use of AHSS continues to grow, there will be an increasingly critical role for forming fluids and lubricants in coming years (Liljengren *et al.*, 2008).

5. CREATIVITY and INNOVATION

Creativity is a process of developing and expressing novel ideas that are likely to be useful. There are four important features of this definition. First, creativity involves divergent thinking, a break away from familiar established ways of seeing and doing. Divergent thinking produces ideas that are novel. Second, these novel ideas must be expressed or communicated to others. This expression provides a reality check on whether the ideas are really novel; or simply bizarre. Third, creativity must also include convergent thinking, some agreement that one or more of the novel ideas is worth pursuing. Fourth, this agreed on option must have the potential for being useful, for addressing the problem that initiated the development of options. The end result of creative process is an innovation. Creativity can be encouraged, enhanced and enabled by managers. Although so much innovation today emerges through 'group' processes, most literature on creativity focusses on creative individuals. Any group can be more creative, even its members individually wouldn't score highly on tests for creativity.

Innovation is the embodiment, combination, and/or synthesis of knowledge in novel, relevant, valued new products, processes, or services. The task to be accomplished, the scope of the desired innovation, dictates the amount of creativity needed. Routine tasks and well defined, well-understood problems may require modicum strength, whereas novel situations and problems require maximum strength. Whether you lead a group of three in a non-profit foundation or 30,000 in a Fortune 500 business, the basic 'process of creativity' is the same. The scale differs, the societal implications differ, and people differ from each other. However, the process derives from some basic interactions among the members of our quirky, infinitely variable, but at the same time surprisingly predictable species. As this book is especially focused on groups as the creating body including: Teams (self-managed or otherwise), Task forces, and

Councils; they all share some essential characteristics included in the following definition. "A group may be thought of as two or more people, existing in an arrangement that permits dome degree of interaction, and sharing some sense of identity as members." However, before the group is launched on a creative voyage, they have to be provided a clear picture of opportunity of innovation; the trigger for creativity. At Beta plant it was a threat that required immediate response to survive. Members of the Beta plant's creative group came from diverse background that at times could 'blur the boundaries' of their disciplines (Leonard & Swap, 2005).

Innovation is also explained as the creation of better or more effective products, processes, services, technologies, or ideas that are accepted by markets, governments, and society. It involves deliberate application of information, imagination, and initiative in deriving greater or different value from resources, and encompasses all processes by which new ideas are generated and converted into useful products. In business, innovation often results from the application of a scientific or technical idea in decreasing the gap between the needs or expectations of the customers and the performance of a company's products.

Purposeful, systematic innovation begins with the analysis of the opportunities. It begins with thinking through the sources of innovative opportunities: Innovation is both conceptual and perceptual; successful innovators look at figures, they look at people and they work out analytically what the innovation has to be to satisfy an opportunity. And then they go out and look at the customers, the users, to see what their expectations, their values, their needs are. An innovation, to be effective, has to be simple and it has to be focused. It should do only one thing, otherwise, it confuses. All effective innovations are breathtakingly simple. Indeed, the greatest praise an innovation can receive is for people to say: 'This is obvious; why didn't I think of it' (Drucker *et al.,* 2008).

17

The role of innovation as a crucial driving force in economic development is widely acknowledged. Within the business setting, innovation is often considered a vital source of strategic change, by means of which a firm generates various positive outcomes including a sustained competitive advantage (Salavou, 2004). This is especially true for incremental (as opposed to radical) innovation, which is vital to increasing a firm's market share and its ability to survive in the industry and leverages a firm's existing resources and capabilities. Within this context an object, service, or process is modified to enhance performance and the recent history of the automotive sector is bursting with examples such as ABS or the Auto Stop Start function. Both are brilliant incremental innovations that were able to evolve in a stable conceptual space employing techniques, materials, and architectural considerations that were already well understood. However, in an economic climate of constant transformation at ever increasing speed, the classical model of innovation, incremental by design, no longer ensures survival against the menace of obsolescence. As new competitors arrive on the market and constantly raise the stakes in terms of quality the continued existence of an enterprise rests on its ability to create ruptures with the past and move forward in a constant state of revolutionary innovation.

A growing body of literature, focusing on different industrial sectors and both large and small firms, regards collaborative practices as a viable method of knowledge creation and transfer. Thus, innovation generation has increasingly become recognized as an outcome of the relationship between a firm and outside entities. Recently, knowledge-sharing with suppliers has received increasing research attention. Manufacturers have discovered the managerial, technological, and financial benefits that may accrue as a result of embedded ties with suppliers. Scholars also generally agree that a substantial part of innovation process occurs between buyers and sellers in the supply chain. Accordingly, a large body of strategy-level research on buyer-seller interaction and innovation outcomes has emerged (Pryor, 2008; Roy *et al.,* 2004).

Industry Collaboration - a new era of open innovation: The move to open innovation has come as a result of major advances in technology and society, which in turn have facilitated the dissemination of information through mechanisms such as the Internet. Today, information can be transferred so easily that it seems impossible to prevent. Thus, the open innovation model states that since firms cannot stop this phenomenon, they must learn to take advantage of it.

As Henry Chesbrough, who coined the term 'open innovation' describes in his book Open Innovation: Researching a New Paradigm: 'Open innovation is the use of purposive inflows and outflows of knowledge to accelerate internal innovation, and expand the markets for external use of innovation, respectively'. Open innovation, however, is a much broader idea. It is not merely a matter of pooling patents, as is the case with open source, but instead has risen organically in company culture, as a result of changes to the global marketplace. Its central idea is that in a world of widely distributed knowledge and fast-moving innovation, companies can no longer rely entirely on their own R&D efforts to ensure market dominance. Instead they need to buy or license processes, patents or trademarks from other companies, as well as to sell or license their own internal IP Rights for profit. Most importantly, they also need to share invention processes and expenditures to push forward the frontiers of knowledge and innovation, while fulfilling consumer needs for every-evolving and interoperable technologies. As a result of all of this, companies have started to look for other ways to increase the efficiency and effectiveness of their innovation processes, whether that be by actively searching for new technologies and ideas outside of the firm or through cooperation with suppliers and competitors, in order to create customer value. Open innovation enables companies to be able to respond in a quick and flexible way to changes in the environment and to remain competitive despite the shortening time-to-market and life cycles of products and technologies. In today's intensely competitive environment, open source business models and collaborative approaches to innovation and business growth are moving beyond 'nice to haves' to 'must haves' (Chesbrough *et al.*, 2008).

There isn't a business that doesn't want to be more creative in its thinking. An established company which, in an age demanding innovation, is not capable of innovation is doomed to decline and face extinction (Smith, 2005). Innovations had better be capable of being started small, requiring at first little money, few people, and only a small limited market (Drucker, 2007). The single most important factor, to igniting creativity, joy, trust, and productivity among employees is a sense of making progress on meaningful work. However, creating an environment that fosters progress necessitates a careful effort. Despite resource-constraints an organization managers can help employees see the meaning in their work. People have to understand what they're doing and why; it's important that the goals be reachable in a realistic time frame - owing to the idea of small wins. People need to know what goal they're trying to reach, but they have to have autonomy in order to get there; it's a delicate balance. People should understand what their mission is. However, micromanaging them shuts down their creative thinking and the value of their unique talents, expertise, and perspectives is lost (Nobel, 2011).

Steady incremental innovations made by employees every day give an organization the sustained growth it needs. Sustained innovation comes from developing a collective sense of purpose, from unleashing the creativity of people throughout the organization and from teaching them how to recognize unconventional opportunities. As innovative ideas surface, a clear sense of mission empowers front-line employees to act on new ideas that further your company's purpose. Leaders create the psychological environment that fosters sustained innovation at all levels. The challenge is that as an organization grows; however, management structures and bureaucracies, designed to channel growth, tend to create barriers to small-scale enhancements. Open communication between management and employees sets the stage for an atmosphere of trust. But if you want to establish a new, more trusting culture, you can't expect employees to take the first step.

Company leadership initiates the process of open communication by sharing information with employees on a regular basis. This includes good news and bad. While larger organizations are often considered less entrepreneurial and inventive than their smaller counterparts, it's not the size of your company that inhibits innovation; it's the systems.

Companies stumble for many reasons: bureaucracy, arrogance, tired executives, poor planning, short-term investment horizons, inadequate skills and resources, and bad luck. Bureaucracy slows down action and impedes innovation. Faster implementation encourages further inventive thinking. An ownership mentality creates a powerful incentive for inventive thinking. When an individual is clearly aware of how his or her interests are aligned with those of the company, he or she has a strong reason to "go the extra mile" to further the mission. When employees don't see how their individual efforts affect company profitability, they tend to be passive and reactive. To encourage greater involvement, make sure each employee knows how his or her work affects company performance. While financial rewards are often tied to innovations, rewarding only the individual or team responsible for the "big idea" or its implementation, sets up a subtle competitive atmosphere that discourages the smaller, less dramatic improvements. Even team-based compensation can be counterproductive if teams are set up to compete with each other for rewards. These incentives discourage the cross functional collaboration so critical to maximum performance (Christensen, 1997).

Companies that successfully foster an innovation culture design rewards that reinforce the culture they want to establish. Tolerating a certain degree of failure as a necessary part of growth is an important part of encouraging innovation. Innovation is a risk. Employees won't take risks unless they understand goals clearly, have a clear but flexible framework in which to operate and understand that failures are recognized as simply steps in the learning process. As your organization innovates you need to practice what Peter Drucker calls "creative abandonment." Projects and processes that no longer contribute should be abandoned to make room for new, progressive activities. Innovation requires optimism. It's about an attitude of continually reaching for higher performance. You can't expect employees to maintain an optimistic attitude if they feel compelled to continue in activities that are going nowhere (Karlsberg & Adler, 2005).

Innovation in manufacturing covers wide areas including, but not limited to, the introduction of new processes, practices, technology, equipment, and new materials. Businesses, with proactive or reactive approach, could resort to innovation in manufacturing for several reasons. In addition to productivity and quality gains, innovation also results in improved responsiveness to customer demands, lower turnaround times, reduced waste levels and downtime, higher product quality, better designed products, capacity for a wider product range, and streamlined relationships with suppliers and customers (Udhas, 2007).

Opportunities for innovation in manufacturing are based on assessment of successful innovations across multiple sectors in manufacturing. The key types of innovation can be classified under the following categories: Innovation in sourcing; Innovation in manufacturing processes; Management innovation; and Innovation through technology. For Innovation in manufacturing processes, companies can innovate in the way products are developed or manufactured, either within the firm or across the supply chain. Such innovations are termed as 'Process Innovation'.

It is typically aimed at garnering competitive advantage through improved quality, reduced costs or reduced time-to-market. For example, one of the greatest innovations to impact manufacturing in the 20th century was the Assembly Line model for manufacturing cars, developed by Henry Ford. The concept, however, did not change the product, but it significantly and permanently changed the process for manufacturing and delivering the product. Several automotive companies, today, use the collaborative product development to shorten the new product development cycles, in collaboration with Tier suppliers (Udhas, 2007).

At Beta plant, 'Process Innovation' is one of the key areas where innovation is exigent. It pertains, primarily, to finding better or more efficient ways of producing existing products. In a manufacturing process; to be called an innovation, an idea must be replicable at an economical cost and must satisfy a specific need (Smith, 2005); end-user innovation is, by far, the most important and critical.

Managers at Beta plant constantly strived to increase efficiency, implement best practice and deliver increased shareholder value. The best way to create value is to innovate your way ahead of the competition. You need to create temporary monopolies where yours is the only show in town. You can do this by harnessing the creative power of your greatest asset, your people. The goal is to turn them into opportunistic entrepreneurs who are constantly looking for new ways of doing business. To build a truly innovative organization you need to have a vision, a culture and a process of innovation. At Beta plant, the key elements of creating a truly innovative and entrepreneurial atmosphere can be summarized in the following, Sloane's, eight steps: Paint an inspiring vision; Build an open, receptive, questioning culture; Empower people at all levels; Set goals, deadlines and measurements for innovation; Use creativity techniques to generate a large number of ideas; Review, combine, filter and select ideas; Prototype the promising proposals; and Analyse the results and roll-out the successful projects (Sloan, 2003).

Innovation based on process need is perfecting a process that already exists, replacing a link that is weak, or supplying a link that's missing (Drucker *et al.,* 2008). Making innovation a ubiquitous capability in manufacturing is fundamentally a leadership challenge. It needs a tangible organizational infrastructure that makes managers accountable at all levels for driving, facilitating, and embedding the innovation process into every part of the culture. Corporate culture has the most important role in the development of innovation (Tellis, Prahbu & Chandy, 2009). Embedded innovative culture in Magna's workforce afforded unprecedented success to the last 50 years of its superb performance. Beta plant was the vanguard of innovative achievements in Magna's group of companies.

6. THE CHALLENGE: TO MINIMIZE DIE CHANGEOVER TIME

This was an extraordinary goal. Besides, reducing the number of die changeovers, the challenge was to curtail the die changeover time by 75 percent, demitting it to 12 minutes, or lower. The daunting question facing the 'Die Changeover Improvement Steering Committee' (DISCO), and Beta plant employees was "how to reduce the die changeover time by 75%, in two suggested phases, each targeting a reduction of 50 percent, while keeping rest of the plants' business running as usual?" DISCO comprised of professionals from production, quality assurance, engineering & design, tool & die, maintenance, finance, customer relations and sales. For tapping into creativity pool of plant employees, DISCO decided to launch the, tried and tested, suggestion system; it also adopted '50-50', a suggested acronym, for two stage time reduction process targeting 50 percent reduction at each stage. Both the acronyms, 'DISCO' and '50-50', corroborated to be catchy and infatuating. It is essential to choose captivating names for the suggestion system that people, ideas and innovation could be associated with. It should be "smooth on the tongue", not too long and easy to handle in publicity. Employees must have a conceptual understanding of the system and the name linked to it (Buchanan & Badham, 2008).

Inadvertently, DISCO's two phase, 50-50, approach, somewhat, mimicked 'phase-gate process; it is a project management technique in which an initiative or project (e.g., new product development, process improvement, business change) is divided into stages (or phases) separated by gates. At each gate, the continuation of the process is decided by, typically, a manager or a steering committee. The decision is based on the information available at the time, including the business case, risk analysis, and availability of necessary resources including money and personnel with appropriate competencies (Cooper, 1986; Hine & Kapeleris, 2006).

7. SUGGESTION SYSTEM: AN EFFECTIVE TOOL FOR IDEA MANAGEMENT

Creativity is a basic human capability. Employees have ideas regardless of whether or not the environment is conducive but the employee will not submit them if the environment is not seen as supportive (Fairbank and Williams, 2001; Stone, 2008). However, in a civilized society, ideas cannot be forced out of people, they themselves need to volunteer them (Pluskowski, 2002). Suggestion systems primarily consist of administrative procedures and infrastructure for collecting, judging and compensating ideas, which are conceived by the employees of the organization (Van Dijk & Van Den, 2002). In addition, suggestion systems have the capability of being all inclusive by being able to focus on capturing ideas from all workers, and not just ideas from identified few smart ones (Fairbank & Williams, 2001).

It is the creativity of employees that forms a source of new ideas, a starting point for innovations; capitalization, involves the transfer of these ideas into innovation. From a perspective of knowledge development and diffusion in the firm, suggestion systems aim at capturing good ideas, the first part of the 'knowledge brokering cycle' (Hardagon & Sutton, 2000). The importance of feedback to employees must not be underestimated. There is a correlation between the number of suggestions submitted and the time taken to give feedback to the suggestor (Du Plessis & Paine, 2007). Two discouraging elements in the suggestion system are the length of time taken to evaluate a suggestion, and the delay in recognition through, sometimes, poor communication channels. The longer the time to give feedback and recognition, the fewer the suggestions submitted (Nel, 2008).The success of the suggestion system depends in the organisation's commitment and involvement, proper policies, procedures and rules, affective administration and process, objective evaluation of ideas and a fair recognition or rewarding system.

A common aim of a suggestion system is to achieve greater employee involvement which eventually leads to greater tangible benefits. Suggestion system should be integrated with the organisation culture (Crail, 2006; Darragh-Jeromos, 2005; Hamel, 2000).

Organisational culture is the pattern of basic assumptions, values, norms and artefacts (the highest level of cultural awareness) shared by the organisation's members (Waddell, Cummings & Worley, 2007). Corporate culture has the most important role in the development of innovation. It plays an important role in the attitude and behaviour of employees and is also an important consideration for recognition (Tellis, Prahbu & Chandy, 2009; Du Plessis, 2007). At Beta plant employees were afforded encouragement, organizational support and committed resources. These have the most direct influence on idea extraction, idea landing, and idea follow up, as avouched by the 'Creativity Transformational Model' which encompasses the main factors that influence the functioning of suggestion systems (Dijk & Ende, 2002).

Suggestion system guidelines were provided to the employees that aimed to answer the six fundamental questions: why were the creative ideas being asked for; what type of ideas were being looked for; who could submit the ideas; how to submit the ideas; how the ideas would be evaluated; and what would happen when the ideas are accepted. Employees were made aware of the award system for accepted suggestions. Recognition and rewards are linked to the psychological contract. Employees want to be recognised or awarded for their efforts and achievements under the psychological contract, if their suggestions are accepted (Holland, Sheehan, Donohue & Pyman, 2007). It is through involving various stakeholder constituencies from the onset of the initiative, creating ideas, the pre-implementation stage, and during the diagnoses of generating ideas that psychological ownership for the suggestion program is established (Van Tonder, 2006). Methods to generate ideas should be clear, straight forward and open to all participants, teams and individuals.

Press operators along other production employees were encouraged to consult maintenance and tooling technicians to ensure technical viability of their ideas. This provided them the much needed assurance that their ideas were functional and doable. Suggestion Boxes were kept available at strategic locations for four weeks and every employee had easy access to them. Suggestion boxes were placed in production facilities, corridors and cafeteria, providing a cost-effective means of collecting paper-based suggestions. Their importance as a tool for collecting suggestions is immeasurable (Du Plessis & Marx, 2009). Suggestion boxes allow a wide range of employees to make their contributions, especially if they do not have access to computers. Despite four weeks of cut off time for submitting, suggestions deposited by the employees were collected every week. DISCO, at their weekly meetings evaluated the suggestions, the posted results showed which suggestions had merit for further assessment and eventual implementation. Some of the suggestions did not directly impact the die changeover time, nevertheless, helped to make Beta plant's manufacturing operations proficient and safer. Toyota launched their 'Creative Idea Suggestion System' in 1951. It was largely a copy of suggestion systems that were in place in U.S. companies at the time, namely the Ford Motor Company. Toyota made some notable innovations to it over the years, but most importantly, they stuck to it. The suggestion system is one aspect of a Lean management system and is embedded in the TPM axioms.

Suggestion Analysis and Implementation - A Short Cut via Experimentation: Senge, De Bono, Basadur, IDEO, Christenson's 'Strategos' and Hamel's 'Innosight', each supplies a distinct component of the operating system for innovation; thinking tools, work practices, culture, market analysis, strategy, education, training and knowledge management. However, experimentation, simulation, discovering options, evaluating alternatives and problem solving, all these exist at the heart of innovation in virtually every discipline (Smith, 2005).

At DISCO meetings, the innovation process, was a blend of methodology, work practice, culture and infrastructure, and analysis method that was guided by five maxims; one conversation at a time, stay focused on the topic, encourage wild ideas, defer judgment, and build on the ideas of others. Beta plant innovators solved problems by focussing upon the useful parameters of a system that, if increased, would enhance it substantially, but also, the harmful aspects that, if left unchecked, would lead to a contradiction. Contradictions are significant, for if eradicated or reduced, directly or indirectly, they contribute to the development of a breakthrough solution. Avoiding compromise is central to innovation. Tradeoffs; strength versus weight, reliability versus cost, service quality versus resource and output versus input, are not the same as an inventive solution that creates new value. Inventive solutions emerge by exploiting useful effects and eliminating harmful effects (Smith, 2005).

Problem solving is a generic skill and can be applied across many different domains. Teams solve problems with science guided experiments that lead to valuable innovations using systematic methodologies. Examples include: the Theory of Constraints (TOC); Critical Chain; Design for Six Sigma (DFSS); Quality Function Deployment (QFD); and the Taguchi Method. To these plethora of strategies is now added something that may be a way of thinking, a set of tools, a methodology, a process, a theory or even possibly a deep science, but which may be gradually shaping up as 'the next big thing.' It's called TRIZ, pronounced 'trees' and is an acronym for the Russian words that translate as "The Theory of Inventive Problem Solving" (Smith, 2005). At Beta plant TRIZ axioms are, ubiquitously, applied in daily operations. TRIZ is the brainchild of Russian scientist and engineer Genrich Altshuller. In TRIZ, learning on the job is a good thing, not a shortcut. TRIZ cannot be studied in any meaningful way unless it is applied to solve problem. To get a sense of TRIZ, think of the theory of constraints but taken to the extreme. TRIZ has been applied in the solution of thousands of such problems, from improving truck fenders at Ford to optimal planning of complex production lines and processes in the oil and fuel industry at AMOCO (Smith, 2005).

DISCO's preliminary meetings availed of de Bono's 'lateral thinking' and adopted his 'Six Thinking Hats' strategy. The term is also used to describe the tool for group discussion and individual thinking. Each hat has a different meaning. Combined with the idea of parallel thinking which is associated with it, the thinking 'Hat' tool provides a means for groups to think together more effectively, and a means to plan thinking processes in a detailed and cohesive way (Birdi, 2005). Because everyone is focused on a particular approach at any one time, the group tends to be more collaborative than if one person is reacting emotionally (Red Hat) while another person is trying to be objective (White Hat) and still another person is being critical of the points which emerge from the discussion (Black Hat). Using a variety of approaches within thinking and problem solving allows the issue to be addressed from a variety of angles (de Bono, 1985).

It is not possible to manage what you cannot control and you cannot control what you cannot measure (Drucker, 1994). Preliminary die changeover stop-watch data revealed that the average time being taken for die changeover was 50 minutes, time on hand between last-part off to first-part off the Press. Bulk of the time was spent in removal of the old coil, setting up the new, and getting the first part off Press. Rest of the time was spent swapping and securing the die and setting up lube lines and spray nozzles. DISCO decided to implement its, two phase '50-50' improvement plan. Workable suggestions were specifically articulated, for the 1st phase, in three areas considering time expended; material coil set up, die swap, and lube lines connections. 1st phase '50-50' target was to bring the die changeover time to 25 minutes. Details, of the amount of resources and time expended at each step involving die changeover were lucidly recorded. This was in addition to the auto-recording of working of the Press and was critical in finding out at what stage the operation was in. Apart from four weeks dedicated to suggestion collection and selection process; DISCO allocated four weeks, each, for 1st and 2nd phase. This allocation included one week, each, for phase end assessment. Entire project was spread over 12 weeks. First and foremost, Job runs were consolidated and with successful negotiations with OEM customers, the number of die changeovers was curtailed to 7-8 per day, three shifts. Focus now shifted to reduce die changeover time, synchronic on all three segments; coil change, die swap, and lube line settings.

8. REALIZING 1ST PHASE - PROGRESSION

With last part off the Press, lube lines were disconnected; carrying web was severed from the partial coil strip, dislodged from the die, and moved to the nearby scrap bin. Operator, using standby 30 ton crane, unloaded the partial coil and placed it in the coil loading/unloading spot to be strapped. Picked up the new coil and loaded it onto coil dispenser's mandrel, ready to be fed to the die for next run. Strapped partial coil was then dropped in the coil bay for storage, and the coil for the next run was brought in and placed in the coil loading/unloading spot. Die was closed - 'shut'; newly installed 'quick' hydraulic die clamps helped to isolate the die from the Press and bolster was ready to be rolled out. Operators, pushed the operating platform out, used new, hand held, 'Magnetic Sweepers' to clear the bolster rails of metal scrap and rolled the bolster out. Bolster weight would flatten, thin, the scrap slugs, if left on the rail, altering the shut height in the next run. Though removal of the last run die, from the bolster, did not affect the die changeover time, nevertheless, afforded more time for next die inspection. New, 'Hydraulic Roller Die Lifters' facilitated die lift. These lifters required a force of approximately one percent of die weight to move the die.

While the north bolster was being rolled out, the south bolster, with new die in place, rolled in. Embedded lube lines were already connected to the lower part of the 'quick connect' lube line distribution block; ready to be coupled with the upper part of the 'block' that held the main lube supply lines. The centerline of the pilots in the progressive die runs parallel with the coil feed direction. To ensure smooth feed, tooling technicians ensured that the edge of the new coil was in alignment with the side rail of the die. Precision key stock, which fit in both the die and bolster, are used for die alignment; the die must be parallel to the coil feeder. These activities were carried out in the pre-staging time. A dedicated die-setter along a tooling technician worked on the die. Die was aligned with the bolster, flexible lube lines adjusted and connectors verified, and the Snap-on lube block was pre-tested for hook-up. A, new, custom made pressure washer was installed in the mobile die wash station. It ensured that a washed, clean, die was available for transfer to the bolster; part of pre-staging operation.

Die and Feed Setup: Most damage to progressive dies takes place during the die setup process: Half-cuts and -forms are made, causing the die to misalign and shear; start-up scrap often is left in the die, causing double metal to be introduced into the tool; pressure, stripper, and draw pads often are half-loaded and unbalanced. The list goes on and on. Establishing a good progressive-die setup procedure is critical to a stamping company's success. Progressive dies often are manufactured to tolerances less than 0.0005 an inch. For the die to function properly, both the top and bottom of the die must be cleaned and freed of debris such as dirt, slugs, grease, and oil. In addition, both the bolster plate and ram of the press must be cleaned. Running a flat file along the bottom and the top of the die removes any high spots that may have resulted from previous dirty die setup.

Die Alignment: The centerline of the pilots in the progressive dies must run parallel with the coil feed direction. It also is important to make sure that the edge of the coil is in alignment with the side rail of the die. Incorrect setting of these items can result in an inability to feed the die smoothly (Figure 6).

Proper die alignment can be ensured in several ways. One is to key the die to the bolster plate. Small blocks of precisely machined steel called keys are fit in both the die and bolster keyways, aligning the die: keyways are precision square cut-outs that are machined into both ends of the lower die shoe and the bolster plate. Other methods are positive stops and locator pins (Figure 7). Regardless of the method used, the die must be parallel to the coil feeder.

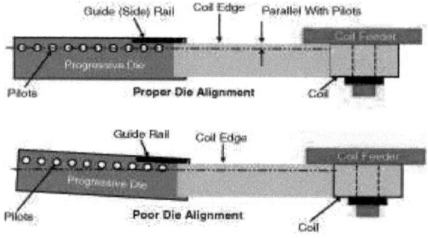
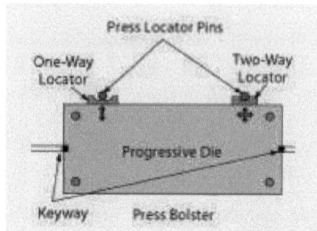

Figure: 6 Proper Die Alignment

Figure: 7 Positive Stops & Locators

Setting Feed Line Height - To keep the feeding material straight and flat, tooling technicians ensured the feed line height was set correctly for the die's feeding level and material was properly fed up to the correct first-hit line (Figure: 8).

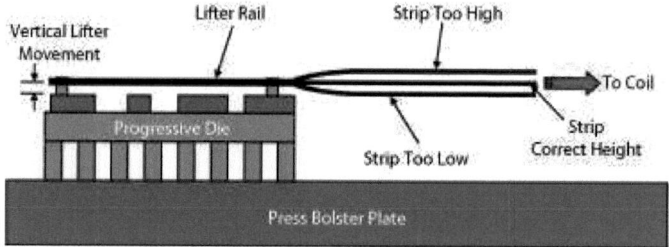

Figure: 8 Correct Strip Height

Setting up the feed line height correctly helps keep the material straight and flat. Straight, flat material has stiffness and is less likely to buckle during feeding; it is critical as most die damage occurs when material is introduced into the die. Starting strip in the wrong location could result in half-cut or -formed parts or unnecessary loose scrap; cutting less than half of a hole or forming less than half of a part will result in unbalanced forces in the tool, resulting in poor die alignment, shearing, or severe die damage. Both the lower and upper dies, after each hit and progression, were inspected and all loose scrap was manually removed. Small pieces of scrap often have a tendency to stick to the faces of cutting punches, pads, and strippers; failing to remove loose scrap in the die results in double metal thickness being fed through it causing severe die damage.

After setting the feed line height, tooling technicians set the 'Pilot Release'. The pilot release function on a coil feeder allows the strip being fed into the die to be released so that the pilots in the die can properly locate and register the strip in the die. For the pilots to position the strip correctly, the feed rollers must unclamp the strip before full pilot entry. The feed release must be timed so that the bullet nose of the pilot enters the strip but the full pilot diameter does not. When this point is established, degrees on the press's stroke are noted and the pilot release is set to let go of the strip at this point. Pilot release is programmed or adjusted so that the material or strip remains fully unclamped until all of the work has been performed in the die and the strip has been brought back up to the feed line height. When the strip is at feeding level, the feed rollers can clamp the strip and feed it forward one progression. This procedure is critical, especially for progressive dies that are making deep-drawn parts and require a great deal of vertical lift to feed forward. Incorrect setup of the feed release can cause: miss-feed; elongation of the pilot holes in the strip; bent, broken, or galled pilots; and poor part location and gauging.

Final Shut Height Calibration: With the die fully loaded, final press's shut height was calibrated, primarily because presses deflect when loaded with a great deal of force. Tooling technicians, at this stage, ensured all scrap removal methods such as slug belts and shakers were in place and functioning properly; gas springs and manifolds were fully charged to the recommended level; all lubrication methods were in place and functioning properly; and die inspected before and after running - check for loose dowels, screws, and debris. Press was run manually to get first part off, part specs were verified and the tooling and quality assurance technicians' signed off the start of production run. Newly installed, better die cushions, aided the process.

Overall time for each die changeover was recorded over three weeks. Individual and overlapping die changeover activities were also timed and data was logged in computer for final tally. Plant employees were rewarded, die changeover time averaged 20 minutes; better than the set target of 25 minutes. However, bulk of the time was still being taken by the coil change and lube lines set-up. It was time to, critically, assess the 1st phase accomplishments and move to the 2nd.

9. REALIZING 2ND PHASE: ACTUALIZATION

While 1st phase implementation was in progress, 2nd phase prelusive remodeling activities set the stage for 2nd phase negotiation. Newly, enlarged 'coil holding area' was re-arranged in two parts purveying to crane capacities; heavy coils were pre-staged in the western side of the coil bay, closest to the Press feeding line, particularly for the higher ranges of strip thickness. Adjustable metal strip bed, equipped with quick clamps, helped secure the strip. Newly installed, ergonomic, laser guided, pneumatic nibbler, moved smoothly to afford a square cut. Its travel was well supported by the overhead guide rail and suspended air-hose caddy. Square cut strip coil was easier to strap, and feed the die in its next run, a time saver. Slug pulling problem was revisited. In case the slug pulls out of the matrix (button) and falls off the punch face, it might cause double metal to be introduced into the die, resulting in die shearing, broken punches, broken die steel, surface defects, and numerous other problems (Hedrick, 2004). Die shoe scrap holes and scrap removal chutes were modified. Occasionally, if the scrap holes or chutes' missed the operators' unclogging, the slugs, from the scrap cumulus, would slip into the tool causing double metal.

Lubrication in Sheet Metal Forming: Lubrication is essential in metal stamping to avoid metal to metal contact and provide cooling effect by preventing heat generation and transfer during deformation. Friction and lubrication in sheet metal forming are influenced by various parameters such as material properties, surface finish, temperature, sliding velocity, contact pressure, and lubricant characteristics (Kim, 2009; *Bay et al.,* 2010). A good understanding of the parameters that affect friction is essential for selecting lubricants and producing good quality sheet metal parts. In sheet metal forming, the magnitude and distribution of friction affect metal flow, part defects and quality, as well as tool wear and production costs; it is one of the process variables that profoundly affect the quality of stamping sheet materials and die changeover time. Using a good lubricant can significantly reduce scrap rate and/or improve the quality of stamping. Four lubricant chemical families, commonly, are used in stamping, and hundreds of formulations are available within each chemical family. Generally, the lubricant families can be rated from the heaviest duty (as far as protecting tooling) to the lightest duty as: Compounded oils (heaviest duty); Macro-emulsions (soluble oils); Chemical solutions (synthetics); and Vanishing oils (lightest duty). The order of this list becomes inverted when these chemical families are rated for ease of use in manufacturing. Tool wear problems in standard metal forming operations of common metals can be addressed by using a heavier-duty product within the same chemical family or switching to a heavier-duty family, such as from a macro-emulsion to a compounded oil. Various types of additives are used to enhance the performance of lubricants. The extreme pressure (EP) additives are very commonly used in heavy-duty metal stamping operations. They are categorized in two types: first - temperature activated; and second non-temperature activated. The temperature activated EP additives such as chlorine, phosphorus, and sulfur react as the interface temperature increases and they generate a film by a chemical reaction with the metal surface. This chemical film helps to prevent metal-to-metal contact in stamping operation. EP additives have different effectiveness in particular temperature ranges such as: Phosphorus is effective up to 205°C (400°F); Chlorine is effective between 205~700°C (1100°F); and Sulfur is effective between 700~960°C (1800°F) (Byers, 2006; Lowery, 2008).

The use of advanced high-strength steel (AHSS) has been steadily increasing in vehicle body constructions to enable improved fuel economy and vehicle safety performance. However, the high press loads required for stamping AHSS sheets causes severe chipping and adhesive wear on stamping/blanking die materials (Wang *et al.,* 2013). A lubricant's function is to minimize contact between the tooling and the work piece. This results in reduced tonnage requirements, longer tooling life, and improved product quality. Lubricant properties, process of application, and control affect tool wear in stamping operations. Most metal forming operations use lubricants to protect the tooling and part from excessive wear caused by scuffing, scratching, scoring, welding, and galling; this ensures quality stampings and reduction I die changeover time (ASTM, 2006; Bergstrm *et al.,* 2008).

Lubricant Application and Control: Even the finest lubricant will not help produce quality part and prevent tool wear if it does not get to the tool when needed. Lubricants can be applied by manual, spraying, immersion or roller depending on the type of the lubricant. Applying emulsion may not be proper with roller because of a separation problem. Spraying is more suitable for much type of lubricants, economical and easy to clean when it needs to be applied locally.

The physical characteristics of the lubricant and metal forming operation determine the application method to be used. Five commonly used application systems are: Roll coater; Drip; Airless spray; Micro-metering mist; and Re-circulating flood. Tool wear can be reduced in tough transfer press operations that use water-soluble flood systems by applying with a roll coater or by drip method a high-viscosity, EP-type compounded oil to the stock going into the die. The compounded oil reduces tool wear on the draw rings in the cupping stations.

An easy and effective way to increase tool life on punching and perforating equipment is to lubricate the bottom of the strip. Punch wear occurs on the return stroke, after the punch breaks through and the metal springs back onto the punch. Using a roll coater or spray system to apply lubricant on the bottom of the strip allows the punch to re-lubricate itself for the return trip. Airless spray systems are suitable for applying lubricants onto progressive dies or transfer tools. They can be actuated to apply lubricant to the top or bottom of the stock or onto only the station or tool needing it. This application method helps to increase the life of extruding punches, sizing tooling, shaving punches, and in-die taps.

During the die changeover project at Beta plant a separate team was concurrently working to come up with an automated lube supply system that would incorporate appropriate methods of dispensing lube with precision. This project was timed to help the reduction in die changeover time as a considerable amount of time was being spent in removing the lube lines from the old die and connecting them to the new.

The lube dispensing system team designed two central mixing and supply stations to deliver pre-mixed lubricant to all of presses. The entire Beta plant's stamping presses used the same lubricant—a synthetic mix with water in two different compositions—one mix for light jobs and the other for heavy. Before installing the lube dispensing unit equipment, press operators would have to fill reservoirs at each press, fed by the central supply stations; supply quality was compromised as the operators would temper with the mix ratio at their will to their liking. Now the supply stations were feeding directly to the presses, one less procedure the operators must worry about; which also helped ensure that the dies consistently received well-lubricated parts and, therefore, won't prematurely wear. Overall, very little attention now was required of press operators to the lubrication process and equipment and time it took for transferring lube lines from one die to another, during die changeover, was considerably reduced.

New lube system allowed the lubricating systems to be set up in a matter of minutes on the press. Nozzles were located on the press with strong magnets and repositioned when a new die was placed in the press; some were built into the die with in-die nozzles and built in lube lines; this took care of the lube supply to the tight confines of the press. Quick disconnects were used on the supply lines so that when a die was to be removed the lines were disconnected and the die removed. When the new die with in-die lube lines was inserted in the press the quick disconnect was reconnected and lube system was ready to go. The lubrication system was connected to the Press control to trigger a press shutdown should the lubricant reservoir run low.

The 120 psi max (average operating pressure under 50 psi) pressure air assist Lubricant Dispensing System dispenses lubricant into a low pressure steam of air or can dispense a controlled drop of fluid, void of air, to the work area. The low pressure air carries the lubricant to the work area and delivers it as a Minimum Quantity Lubrication non-fogging spray. The dispensing of the lubricant can be controlled with programmable Electronic Controllers and Air Timers.

The system controls allows the operator to shut off nozzles that are not needed for a particular die set up. The operator then had the ability to individually program each of the operating nozzles as to when they will spray and how much lubricant each nozzle will dispense during the cycle of the press. The system can dispense lube in predetermined repeatable actuations, single shots or be tied into a machine and dispense on each cycle of the machine. The new lube system; reduced Fluid Consumption, provided near dry Lubricating through non Fogging Spray, reduced cost of parts clean-up, and saved on environmental cost of disposing waste lube mix. The new lube dispensing system consistently and reliably delivered lubricant to the die and onto the work-piece; beside production of quality stampings it helped to cool and extend tool life.

Steel tubing lube lines were embedded in all dies. Half nozzles were installed in the confined die spaces and snap on lube distribution blocks were installed at the lube lines die junction. To address its lubrication concerns in the plant, the recently inaugurated 'Central Lube System' (CLS) was connected to press line. CLS ensured dies consistently receive well-lubricated parts and, therefore, won't prematurely wear. Programmable controllers were modified to include individual nozzle spray quantity control. Newly installed lube low pressure and low level alarm was connected to maintenance control desk/dispatch. With lube process automated, operator involvement was reduced to minimal. Downtime was reduced by 20%, for heavy gauge material. Overall, very little attention now was required of plant operators to the lubrication process and equipment. It is imperative to have dependable, repeatable lubricant application.

Beta plant stamped as much as 1 million lb. of steel per week. Press at augmented line speed of 18 strokes per minute (SPM), at times, generated a lot of heat. Any lubrication failings could quickly lead to galling and other premature die-wear issues caused by overheating. However, experimentation of DC-53 tool steel, instead of D-2 for arduous die sections, tremendously improved die performance. Hourly consistency check of lube mix was also included in the modified stamping procedure and a Lube boundary film monitoring process was set up for critical points. CLS afforded 30% savings in lube cost (bought less) and substantial savings in disposal cost of 11cents per liter (environmental cost). DISCO team discovered seepage into the waste lube underground pit, rain water was getting in. This was discovered while monitoring the quantities consumed, there were noticeable dissimilarities among volumes in and out.

Projecting 2nd phase prelusive remodeling activities, a modified die-set up procedure was introduced that ensured no half-cuts and -forms were made causing the die to misalign and shear, there was no start-up scrap left in the die causing double metal to be introduced into the tool, and no pressure, stripper, and draw pads were half-loaded and/or unbalanced. Parallel activities pertaining to modification/fabrication of scrap chutes, installing lube lines and spray nozzles, did not subvert the die changeover time window. Given the time constraints, some of the work on dies was outsourced. Stamping job runs were assessed for quantity and frequency on JIT basis

42

and each die was allocated a specific, marked, floor space. JIT bins arrival was re-organized. Bins, in which the Beta plant sent its parts to customers, were sent back; bins were customers' property.

Most needed spare parts were stored in proximity to the Press. Jigs for spec check were moved next to the Press work station. Vertical carousel storage for each die was sorted out and re-arranged. All pneumatic appliances and hoses for emergency use were lifted off the floor area surrounding Press and elevated, retractable caddies were installed. Light weight pendent control units with boots were provided for Press operation. Newly installed laser pointers to position the die on bolster locator pins cinched the die setting operation. Light curtains were isolated from the Press frame, lowering set up time. Reduced number of die changeovers extended run time and provided sufficient time to the die-setters and tooling techs for efficiently pre-staging die for the next run. It gave them ample time for inspection of: die plates (foundation for mounting die components); guide pins and bushings (align the upper and lower die shoe); heel blocks (contain wear plates to absorb side thrust); screws, dowels, and keys (to fasten and locate die components); stripper (pull metal off cutting punches); pressure (hold down metal in wipe bending process); draw pads (control metal flow during the drawing process); spools, shoulder bolts, and keepers (to fasten pads to the die shoes while allowing them to move up and down); retainers (hold the cutting-forming components' to upper and lower die plates); and springs (coil, urethane, or gas - supply the force needed to hold, strip, or form metal). Adjustments that emanated from these inspections, not only reduced, considerably, the time taken for 'first part off Press'; it tremendously improved the, overall, dies' performance. Small incremental gains in time accumulated to substantial reduction in die changeover time.

Catwalk platforms were modified. Slimmer, upper part - the actual walkway, replaced with aluminum alloy expanded metal plates, rendered it considerably lighter while newly installed casters ensured its quick removal. Fitted with new magnetic sweepers, the catwalk while being hauled, picked up scrap pieces from the rails; making the bolster exit smoother and quicker. To ensure safe and secure coil storage, improved 'Roll-Blocks' were installed in the coil holding area for heavy coils. To enhance coil storage capacity, and for close stacking of coils, a motorized, C-hook was installed. C-hook is a, below the crane hook, vertical lifting device that enable operators to handle vertical coils in a safe, efficient, and economical manner. Its automatic latch assembly enhances efficiency by requiring just one person to operate it. With C-hook one operator could now, proficiently, position the heavy metal strip coil on de-coiler. C-hook use replaced slings and reduced end coil scrap. To avoid crane crash in an accidental load swing, beeping 'Proximity Sensors' were installed around north-east and south-east corners of the Press. Considering, 30 ton crane's availability, production runs were planned to accommodate die changes so that crane was not pre-occupied with any other job.

Additional lighting was added to focus on the problematic areas. Non vibrating light fixtures were added to illuminate die stations. Coil pre-feed check alarm was added. Banding apparatus was located close to the coil feed, inside the coil cage; and self-locking, banding strip rolls dispensers were provided. In addition to the new coded buzzer system for, maintenance and tooling, service calls, dedicated tooling and maintenance technicians were at stand by – it reduced service call travel time. All the modifications and alterations paid off. Die changeover time, recorded over three weeks, averaged 10 minutes, right on target. However, bulk of the time was still being taken by the coil change. It was time to, critically, assess accomplishments of both, 1st and 2nd phase, and devise procedures that would help sustain this expeditious die changeover.

10. DENOUEMENT and DISCUSSION

DENOUEMENT: Out of the total 216 suggestions received, 72 were implemented during the entire program. 48 of the suggestions directly induced improvement in the die changeover time, 22 in the 1st and 26 in the 2nd phase; while remaining 24 suggestions, though not directly pertaining to die changeover, enhanced Beta plants' manufacturing operations and improved plants' safety. In DISCO's concluding meeting, all the process of impelling innovative ideas, suggestions system performance, number of suggestions solicited and implemented, were analysed. A workshop of 'lessons learned' was held and procedure for sustaining the die changeover improvements was formalized and adopted. For the future, the process was linked to the 'continuous improvement' strategy. As per Beta plant's rewards policy, all innovators whose suggestions were implemented were rewarded generously and fairly; a token of appreciation was also given to all the participants whose suggestions were not implemented. It is essential for the firm's success to reward innovators for their contributions. At Magna, reward system is an important integrant in managing innovation. Although, appropriate, cultural values and norms are a powerful means of stimulating creativity and innovation; a system of prizes is the best possible mechanism for eliciting innovation, provided the size of the prize could be linked to the social value of the innovation (Price, 2007; Gallini and Scotchmer, 2001).

DISCUSSION: Whole process of effectuating innovations, from idea to implementation, is fraught with applied theoretical underpinnings and TPM axioms. As the story of implementing innovative suggestions progresses the reader would notice inter twined nature of real life situation where multiple theories, models, and frameworks overlap and complement one another; the problem solver/innovator moves seamlessly from one domain to another in pursuit of solution.

Experimentation of tool steel DC-53 for arduous die sections, turned efficacious: because of the steel's ability to withstand compression and shock, die could make up to 5 times more hits, double to triple the tool life, 30% less machining time, much faster grinding time and 30% less downtime when used in progressive dies. The steel sharply reduced cracking and chipping and provided better hardness after heat treatment. Welding and general repair was much easier with DC-53 than with D-2 (Cummings, 2001). This, along with new die set up procedure, and modifications to the coil feed system and lube dispensing system, enabled press to make higher strokes - from 14 to 18 SPM.

Digital imaging of the repairs done by die technicians exposed various skill deficiencies. They needed training. A repair standard was introduced to verify the quality of repair done by the technicians and keep them abreast with the latest developments in die materials, manufacturing, and repair technologies. Consequently die repair procedures were standardized.

Designing the automated lube dispensing system with advanced lubrication technology had been the most supporting program implemented during the die changeover project. With previous oilers, Beta plant's operators constantly fought maintenance battles with the lube lines, nozzles, valves, pumps, filters, etc.; implementing TPM was tried at the lubrication cell level, but the work wasn't getting done to the standard. The lube system design team bypassed operator involvement to let them focus on other more important tasks around their press cells and decided to automate the lubrication process; this, not only, reduced Press downtime due to lubrication issues throughout the plant by at least 20 percent, it also helped to reduce the Press die changeover time considerably. To achieve this, the team at Beta plant engaged in a three way open innovation involved collaboration on two fronts: internally between maintenance, tooling, and engineering departments; and externally, between lube suppliers, lube dispensing equipment makers, and raw material producers. The realized befits included: The ability to leverage R&D developed on someone else's budget; Extending the reach and capability of new ideas and technologies during R&D and M&A processes; The opportunity to refocus internal resources on managing the implementation of new technologies; The ability to conduct strategic experiments at lower levels of risk and resources; and the opportunity to create an innovative culture, from the 'outside in' through continued exposure and relationships with external innovators. This collaboration afforded better quality parts with significant drop in scrap rate and a state of the art lube dispensing system that delivered lube precisely; in time, at the right spot, and in the right quantity.

Collaboration with Beta plant's lubricant supplier, IRMCO, provided a unique opportunity to resolve the 'Work Hardening and Spring Back' problem faced by the plant. Stamping AHSS can push the capability limits of some lubricants and often causes lubricant film break down and galling that increases scrap rates and tool maintenance cost. For this reason, proper lubrication is vital for successful sheet metal forming; it is essential in stamping operations to avoid metal to metal contact and provide cooling effect by preventing heat generation and transfer during deformation. Though AHSS is most advantageous when used for safety components, structural parts of the car body and the chassis, its increasing use puts higher demands on tool steels and lubricants used in forming and blanking/punching operations.

The work-hardening and spring-back characteristics of AHSS can require stamping operations to boost press tonnage and increase ram dwell time. The resulting increase in friction and associated heat can break down stamping lubricants, deplete their boundary protection and render them ineffective. In these situations, stampers opt for lubricants with extreme-pressure (EP) additives such as chlorine, sulphur and phosphorous. Activated by heat, these additives react with metal to form metallic "salts" with low shear strength. These salts provide an additional yet temporary coating to protect the tools and work piece. As the tool and work piece temperatures increase during deformation, lubricants with EP additive become thinner (low viscosity), may burn. On the other hand, the lubricants with extreme temperature (ET) additive become thicker (high viscosity). Lubricants with ET additives stick to the hot work piece and create a friction-reducing film barrier between the tool and the work piece (IRMCO, 2012). Collaboration with IRMCO resulted in production of quality stampings with reduced scrap rate in a short time span that would have taken lot longer otherwise; obviously it helped to reduce the die changeover time.

Managers should know when to compel silence and listen, and dispense the four nourishers necessary for a healthy inner work life; respect and recognition, encouragement, emotional support, and, finally, affiliation. Any action that serves to develop mutual trust, appreciation, and even affection among coworkers is conducive to affiliation. Companies that successfully foster an innovation culture design rewards that reinforce the culture they want to establish. Implementation of ideas creates synergies and opportunities for more ideas. Besides impeding progress, shifting goals can drain work of its meaning; when people perceive that their hard work will not amount to anything, they come to feel that they are wasting their time, and that their work is without value. Goals can shift for many reasons, but the consequences for inner work life are almost always negative (Amabile and Kramer, 2011).

Management cannot deliver the change on its own. The best source for the idea-generation and creativity needed for innovation is the team within the organization. To turn them into entrepreneurs who are ardently looking for new opportunities one has to first empower them. The purpose of empowering people is to enable them to achieve the change through their own efforts. They need clear objectives so that they know what is expected of them. They need to develop the skills for the task. They need to work in cross-departmental teams so that they can create and implement solutions that will work across the organization. They need freedom to succeed. And when you give someone freedom to succeed you also give them freedom to fail (Sloane, 2003).

Sloane further elaborates that "people are anxious about change; it is uncomfortable. Change means winners and losers; it is natural that people will prefer to stay within their comfort zones rather than risk an embarrassing or costly failure. Lower management should spend time with people encouraging them to undertake risks and reassuring them that those risks are necessary and worth taking. Fear of failure often inhibits people from pushing themselves to new limits. Employees need to know that doing nothing has its risks too; that staying in the corporate comfort zone is a dangerous option. They ought to be reassured that they will not be punished for taking risks, for worthwhile failures, for bold initiatives that do not succeed. Of course, taking risks means taking calculated risks not wild risks; every employee who is undertaking a risky initiative needs freedom but needs mentoring and guidance as well."

The process of finding creative solutions is something that can be built into the culture of the organization. This is done by techniques, methods, workshops and a pervading attitude of encouragement for radical ideas. The innovation process involves the generation of many ideas in response to a given issue or challenge. Businesses that are fast to market carry out quick pilot tests rather than spending months in "paralysis by analysis" (Sloane, 2003).

Tried-and-true methods such as suggestion systems that promote cost-saving ideas and continuous improvement task forces that look to make processes more efficient, can certainly improve organization's bottom line. Innovation is not a one-time exercise. It involves continuous efforts in re-inventing the firm's products, services and processes in the light of market and technology developments. Sustained growth and profitability can be achieved through the integration of three critical levels; People, processes and tools (Constantinides, 2012)

11. CONCLUSION

Idea management systems do not replace traditional departments and processes involved in new services, products, or strategies. They serve as an adjunct to them and provide a framework that can help organizations turn innovation into an enterprise-wide discipline-and a sustainable process that drives growth in good times and bad (Tucker, 2003). Ideas are rarely rejected on their merits; they're rejected because of how they make people feel. The bigger the idea, the harder the persuasion challenges. The challenge is ideas don't come with the courage to invest in them. Good ideas are everywhere: what's uncommon is people with the conviction to put their reputation behind ideas (Berkun, 2014).

Projects will always require a significant investment in materials, personnel, and adequate time; keeping projects resource-poor, will impede creativity. Employees will feel inhibited if they don't feel comfortable asking for support or, worse, if they feel that others are deliberately blocking necessary information from them. Solution through innovation is a learning process, from both failures and successes. Managers and co-workers should not punish or ridicule someone who tried and failed. Instead, the experience should be turned into a sense of progress; having learned something.

Innovation is a risk. Employees won't take risks unless they understand goals clearly, have a clear but flexible framework in which to operate and understand that failures are recognized as simply steps in the learning process. It is imperative that management provides environment that encourages curiosity, persistent investigation, and the shared perception that failure in one realm may translate into success in another. Moreover, each invention did take more than one person's initial "aha" to become profitable; such management, thus, provides the resources for groups to explore and follow through (Leaonard & Swap, 2005).

Einstein said "If I had 20 days to solve a problem I would take 19 to define it." There are many creative ways to think about a problem, and different ways to look at a situation. The impatient run at full speed into solving things, speeding right past the insights needed to find a great solution. If you listen to how successful creators talk about their daily work, they spend more time thinking about the problem than epiphany obsessed media would have us believe. How would you feel about an invention that ends your profession? All innovation is change and all change helps some people and hurts others. Any successful idea has a multitude of consequences that are impossible to predict and difficult to even measure (Berkun, 2014).

Beta plant's team constituted primarily of 'blue collar' employees who did not read many books on 'Innovation' in order to innovate. They worked; diligently experimented to find the solutions. It's not clear why anyone should read a book about innovation. There's little evidence people we'd call creative got that way by reading a particular book. Most skills in life are only acquired by work, and to be more creative means to create and learn, rather than merely read. The challenge with creative work, especially in a marketplace, is the many factors beyond ones control. One may do everything right and still fail. Most books on creativity make big promises based on history: they cherry pick examples from the past and claim it's predictive. Methods can be useful but they deny that the present is different from the past. There are too many variables in the present to have certainty. This is why terms like innovation system or innovation pipeline are absurd. The idea of an innovation portfolio, where a range of risk is assumed, is more honest. Many books on creativity are surprisingly uncreative and unreal (Berkun, 2014).

Auto industry is going to remain important, if ecologically sustainable transport systems are to be developed for local, regional and national economies, as well as for the future of the planet; it will, no doubt, remain an important topic in the academic literature (Bailey *et al.*, 2010).

REFERENCES

Advantage, 2014, retrieved on 30th June 2014 from: http://www.advantagefabricatedmetals.com/stamping-process.html

Amabile, T., and Kramer, S. (2011). *The Progress Principle: Using Small Wins to Ignite Joy, Engagement, and Creativity at Work*. Boston: Harvard Business Review Press.

ASTM standard G40 (2006)

Bailey, D., Clark, I., De Ruyter, A., "Private equity and the flight of the phoenix four: the collapse of MG Rover in the UK", Cambridge Journal of Regions, Economy and Society 2010; 3.

Bay, N., Azushima, A., Groche, P., Ishibashi, I., Merklein, M., Morishita, M., Nakamura, T., Scmid, S., Yoshida, M., 2012, "Environmentally Benign Tribo-systems for Metal Forming", CIRP Annals, 59 (2), 2010

Bergstrm,J., Krakhmalev, P., Grd, A., and Lindvall, F., 2008, "Galling in Sheet Metal Forming," in proceedings of the IDDRG 2008.

Berkun, S., 2010, "The Myths of Innovation", O'Reilly Media, CA (USA). ISBN-10: 1449389627

Birdi, K. S., 2005, "Evaluating the effectiveness of creativity training", Journal of European Industrial Training, Vol. 29 No. 2, 2005, pp. 102-111, Emerald Group Publishing Limited

Buchanan, D. A., and Badham, R.J., (2008), *Power, Politics and Organisational Change*. Sage Publications, London.

Byers, J. P., 2006, "Metalworking Fluids" Second Edition, CRC Press, Boca Raton, FL, pp. 104-114

Chesbrough, H. (2003), Open Innovation: The New Imperative for Creating and Profiting from Technology, Harvard Business School Press.

Chesbrough, H., Vanhaverbeke, W., and West, J., eds., Open Innovation: Researching a New Paradigm. Oxford: Oxford University Press, 2006. ISBN: 0-19-929072-5.

Christensen, C., 1997, The Innovator's Dilemma: When New Technology Causes Great Firms to Fail. Boston: Harvard Business School Press.

Clegg, S., Kornberger, M., and Pitsis, T., 2008, "Managing Organisations", Sage, London

Constantinides, C., 2012, "Product innovation: Unifying people, processes and tools", retrieved from: http://www.innovationtools.com/Articles/EnterpriseDetails.asp?a=744

Cooper, R. G., 1986, Winning at New Products, Addison-Wesley. ISBN 978-0-201-13665-4.

Costello, N., Michie, J., Milne, S., 1989, "Beyond the Casino Economy", London: Verso.

Crail, M., 2006, "Fresh Ideas from the Floor", *Personnel Today*, June 20, 2006, pp.30

Cummings, C., 2001, Canadian Machinery and Metalworking, retrieved from: http://www.imsteel.com/pdf/is_it_tough_enough_for_you.pdf

Darragh-Jeromos, P., (2005), "A suggestion system that works for you", *Super Vision,* Vol. 66, Issue 7, p. 18.

David B., de Ruyterb, A., Michiec, J., and Tylerd, P., 2010, "Global restructuring and the auto industry", Cambridge Journal of Regions, Economy and Society, Volume 3, Issue 3, Pp. 311-318

de Bono, Edward (1985). Six Thinking Hats: An Essential Approach to Business Management. Little, Brown, & Company. ISBN 0316177911 (hardback) and 0316178314 (paperback).

Dijk, C. and Ende, J. 2002, "Suggestion systems: transferring employee creativity into practicable ideas", R&D Management 32, 5, 2002, Blackwell Publishers Ltd., MA, USA

DOCOL, 2012, DOCOL's Advanced High Strength Steels for the Automotive Industry, DOCOL-SSAB Automotive Brochure, retrieved on 14 Jan 2014 from: http://www.ssab.com/Global/DOMEXDOCOL/Brochures/en/490_SSAB_Automotive_final.pdf

Drucker, P. F., (1994). The theory of business, *Harvard Business Review, September-October.*

Drucker, P. F., 2007, "Innovation and Entrepreneurship", Butterworth-Heinemann, MA, USA

Drucker, P. F., Collins, J., Kotler, P., Kouzes, J., Rodin, J., Rangan, V. K., et al. (2008). The Five Most Important Questions You Will Ever Ask About your Organization, p. xix (2008), retrievable from: http://info.emeraldinsight.com/drucker/index.htm

Du Plessis, A. J., 2007, "Change, Organisational Development and Culture: Human Resource's Role in a future SA", International Review of Business Research Paper, 3(1), March: 1- 10

Du Plessis, A. J., Pain, S., (2007), Managing of Human Resources and Employment Relations in New Zealand's Retail Industry, *The International Journal of Knowledge, Culture and Change Management,* 7(2): 83-91

Du Plessis, A., and Marx, A., 2009, "Suggestion system as perfect tool for entrepreneurs to be successful in New Zealand: some empirical evidence, ASGE New Zealand

Engelberger, J. F. (1982). Robotics in practice: Future capabilities. Electronic Servicing & Technology magazine.

Fairbank, J.F. and Williams, S.D. "Motivating Creativity & Enhancing Innovation through Employee Suggestion System Technology", Creativity and Innovation Management, 10 (2): 68-74 (2001).

Freyssenet, M., Jetin, B., 2009, "Big Three: Le Piege de la 'Liberalisation' Salariale et Financiere se Referme", La Lettre de GERPISA, January–March, 9–14.

Gallini, N., S. Scotchmer (2001) Intellectual Property: When is it the Best Incentive Mechanism?, Innovation Policy and the Economy, Vol. 2, p. 51-78.

Hamel, G. (2000) Leading the Revolution, Boston MA: Harvard Business School Press

Hardagon, A. and Sutton, R. I. (2000) Building an innovative factory. Harvard Business Review, 78, May-June, 157-166

Hedrick, A., 2004, "Eliminating slug pulling during piercing operations" retrieved from: http://www.thefabricator.com/article/toolanddie/eliminating-slug-pulling-during-piercing-operations

Hine, D., and Kapeleris, J., 2006, "Innovation and Entrepreneurship in Biotechnology, An International Perspective: Concepts, Theories and Cases", Edward Elgar Publishing, p. 225, MA, USA, ISBN 978-1-84376-584-4.

Holland, P., Sheehan, C., Donohue, R., and Pyman, A., 2007, "Contemporary issues and challenges in HRM", Tilde University Press, Australia

Karlsberg, R., and Adler, J., 2005, "Severn strategies for sustained innovation", retrieved from: http://www.innovationtools.com/Articles/EnterpriseDetails.asp?a=185

Kim, H., Palaniswamy, H., and Altan, T., 2006, "Investigation of tribological conditions in forming uncoated and galvanized advanced/ultra-high strength steels, CPF, 2, 3/06/01

Kim, H., 2009, Advanced High Strength Steels (AHSS): Evaluation of lubricants, tool materials and coatings for reducing galling, LAP Lambert Academic Publishing (August 14, 2009), ISBN-10: 3838305906

Klier T., and Rubenstein J., 2010, "The changing geography of North American motor vehicle production", Cambridge Journal of Regions, Economy and Society: 3 (3); 311-318.

Leonard D. A., and Swap, W., 2005, "When Sparks Fly: Harness the Power of Group Creativity", Harvard Business School Press, USA

Lowery, S., 2008, "Green Lubes under Microscope", The Fabricator, October 2008

Magna, 2012, retrieved from: http://www.magna.com/about-magna

MetalForm, 2012, at: http://www.metalformingmagazine.com/magazine/article.asp?aid=5590

Nel, L., 2008, "The usefulness of corporate ethics programmes in integrating ethics into an organisation's culture", unpublished Master of International Communication Thesis, Unitec New Zealand

Nobel, C., 2011, "How Small Wins Unleash Creativity", Harvard Business School, Working Knowledge   Paper Published: September 6, 2011

Pluskowski, B., 2002, "Dynamic Knowledge Systems", White Paper- 0602- 1 Imaginatik, available at: www. imaginatik. Com

Price, R. M. (2007) Infusing innovation into corporate culture, Organizational Dynamics, Vol. 36, no. 3, p. 320-328

Pryor, J., 2008, "Industry Collaboration: A new era of open innovation", retrieved on 8th June 2014 from: http://www.cpaglobal.com/download_centre/white_papers/open_innovation

Roy, S., Sivakumar, K., & Wilkinson, L. F. 2004. Innovation generation in supply chain relationships: A conceptual model and research propositions. Academy of Marketing Science, 32 (1): 61-79.

SSAB, 2013, "Tooling solutions for advanced high strength steels – selection guidelines" SSAB Swedish Steel Brochure, retrieved on 14 January 2014 from: http://www.uddeholm.com/files/Tooling_solutions.pdf

Salavou, H., (2004) "The concept of innovativeness: should we need to focus?", European Journal of Innovation Management, Vol. 7 Iss: 1, pp.33 - 44

Samuelson, P. A., and Nordhaus, W. D., Microeconomics. 17th ed. page 110. McGraw Hill 2001.

Sloane, P., 2003, "Innovation: Creating the best practices of tomorrow", retrieved from: http://www.innovationtools.com/Articles/EnterpriseDetails.asp?a=78

Smith, H., 2005, "What Innovation Is", A CSC White Paper, European Office of Technology and Innovation, retrieved from: http://www.ideationtriz.com/what_innovation_is-v25.pdf http://www.stage-gate.com/knowledge.php

Stanford, J., 2010, "The geography of auto globalization and the politics of auto bailouts", Cambridge Journal of Regions, Economy and Society, (2010) 3 (3): 383-405.

Stone, R. J., 2008, "Human resource management", 6th edition, John Wiley and Sons, Australia Ltd

Tellis, G. J., Prabhu, J. C., Chandy, R. K. (2009) Radical Innovation Across Nations: The Pre-eminence of Corporate Culture, Journal of Marketing, Vol. 73, p. 3–23.

Tucker, R. B., 2003, "Driving Growth Through Innovation: How Leading Firms Are Transforming Their Futures.", retrieved from: http://www.innovationtools.com/Articles/EnterpriseDetails.asp?a=75

UDDEHOLMS, 2013, TOOLING SOLUTIONS FOR ADVANCED HIGH STRENGTH STEELS, retrieved on 14 January 2014 from: http://www.uddeholm.com/files/SB-tooling-solution-ahss-uddeholm.pdf

Udhas, P., 2007, Innovation in Manufacturing, KPMG, India, retrieved from: http://www.kpmg.de/docs/innovation_in-manufacturing.pdf

Van Dijk, C. and Van Den Ende, J., 2002, "Suggestion Systems: Transferring Employee Creativity into Practicable Ideas", R&D Management, 32(5): 387-395.

Van Tonder, C. L., 2006, "Organisational Change", Van Schaik Publishers, Pretoria

Waddell, D. M., Cummings, G. T., and Worley, C. G., 2007, "Organisational Development and Change", Thomas, Melbourne, Australia

Wang, C., Chen, J., Xia, Z. C., and Ren, F. 2013. Die wear prediction by defining three-stage coefficient K for AHSS sheet metal forming process. The International Journal of Advanced Manufacturing Technology. October 2013, Volume 69, Issue 1-4, pp 797-803

Womack, J. P., Jones, D. T., & Roos, D. (1990). The machine that changed the world. New York, NY: Rawson Associates

Printed by Books on Demand GmbH, Norderstedt / Germany